DFG

Amiodarone Profile

Distribution:

VCH Verlagsgesellschaft, P. O. Box 10 11 61, D-6940 Weinheim (Federal Republic of Germany)

Switzerland: VCH Verlags-AG, P. O. Box, CH-4020 Basel (Switzerland)

United Kingdom and Ireland: VCH Publishers (UK) Ltd., 8 Wellington Court, Wellington Street, Cambridge CB1 1HW (England)

USA and Canada: VCH Publishers, Suite 909, 220 East 23rd Street, New York 10010-4606 (USA)

ISBN 3-527-27378-6 (VCH Verlagsgesellschaft)
ISBN 0-89573-963-1 (VCH Publishers)

ISSN 0930-7958

DFG Deutsche Forschungsgemeinschaft

Amiodarone Profile

edited by Robert A. A. Maes
and compiled by David W. Holt, Theo A. Plomp
and William J. McKenna

Report XIII
of the Senate Commission for
Clinical-Toxicological Analysis

Mitteilung XIII
der Senatskommission für
Klinisch-toxikologische Analytik

Deutsche Forschungsgemeinschaft
Kennedyallee 40
D-5300 Bonn 2
Federal Republic of Germany

Telefon: (0228) 885-1

Library of Congress Card-No. 89-7-0679

British Library Cataloguing in Publication Data
Amiodarone profile.
1. Man. Heart. Arrhythmia. Drug therapy. Amiodarone
I. Maes, Robert A. A. II. Deutsche Forschungsgemeinschaft
III. Series 616.128061

ISBN 3-527-27378-6

Deutsche Bibliothek, Cataloguing in Publication Data

Amiodarone Profile / DFG, Dt. Forschungsgemeinschaft. Ed. by Robert A. A. Maes and comp. by David W. Holt ... – Weinheim ; Basel (Switzerland) ; Cambridge ; New York, NY : VCH, 1990
(Mitteilung ... der Kommission für Klinisch-Toxikologische Analytik / DFG, Deutsche Forschungsgemeinschaft ; 13)
ISBN 3-527-27378-6 (Weinheim ...) brosch.
ISBN 0-89573-963-1 (New York) brosch.
NE: Maes, Robert A. A. [Hrsg.]; Deutsche Forschungsgemeinschaft; Deutsche Forschungsgemeinschaft / Kommission für Klinisch-Toxikologische Analytik: Mitteilung ...

Production Manager: L&J Publikations-Service GmbH, D-6940 Weinheim
Composition: Lichtsatz Glaese GmbH, D-6944 Hemsbach
Printing: Zechnersche Buchdruckerei, D-6720 Speyer
Bookbinding: Progressdruck GmbH, D-6720 Speyer

Printed in the Federal Republic of Germany

Contents

Preface

There has been a growing interest in serum drug monitoring over the past decade since it has become obvious that drugs are more often over- or underadministered than was generally considered to be the case. For instance, a retrospective study covering several years found that phenytoin and theophylline concentrations were within the therapeutic range in only about 50% of the cases studied.

A firm knowledge of pharmacokinetics is essential with respect to the determination and the interpretation of drug serum concentrations. It is evident that the pharmacological actions of certain drugs correlate better with their concentrations in serum or plasma than with the doses administered. Retrospective and current studies showed a clinical need for the individualisation of dosage regimens based on drug monitoring, e. g. in the cases of theophylline, tobramycin, antiarrhythmics and some antiepileptics.

Because of the similarities of clinical and toxicological problems and since many techniques are applied to the solution of both, the "Senatskommission für Klinisch-toxikologische Analytik" of the Deutsche Forschungsgemeinschaft started a reappraisal of drug monitoring.

This manuscript is a result of the penetrating discussions within the working group "Drug Monitoring". It appears in a series of publications of the Senate Commission dealing with monitoring of those drugs for which the determination of the concentration in blood is of paramount importance for a state-of-the-art therapy. Profiles on other drugs, e. g. phenytoin, digoxin, methotrexate and cyclosporine will follow.

A book on classic and modern aspects of pharmaco- and toxicokinetics with special regard to drug monitoring has been published by Van Rossum and Maes as a starting point of the whole series (Report IV).

Goals of the Senate Commission

The principal goals of the "Senatskommission für Klinisch-toxikologische Analytik" are to evaluate and compile procedures for the determination of drugs and poisons which are acutely toxic to man. The analytical procedures used should be as simple as possible and preferably designed to be applied at the bed side in hospitals.

The work of the Senate Commission aims at

- the evaluation and improvement of procedures for qualitative and selective detection of poisons,
- the refinement of methods for the estimation of poisons in biological fluids in order to
  - recognise toxic substances,
  - determine the degree of intoxication, and
  - evaluate the results of therapy (i. e. the effectivity of detoxification),
- the evaluation of methods for monitoring therapeutic drugs in hospitalised and out-patients, and
- the collection and evaluation of analytical and clinical data of poisons on the widest possible scale.

Introduction

Measuring antiarrhythmic drugs in plasma is evident and necessary since most of these compounds show a narrow therapeutic range and may demonstrate toxic effects, if no monitoring is performed.

Amiodarone is the first agent with predominantly class III effects that is effective after oral administration.

Several factors, such as serious adverse effects stand up for the measurement of amiodarone and its major metabolite during therapeutic use and are grounds for writing a profile of this antiarrhythmic drug.

In contrast to other monographs in this series we have chosen another framework, effected in three different chapters dealing with the analytical methodology and pharmacokinetics of the drug, its therapeutic monitoring and the management of patients treated with amiodarone.

In our opinion only this plan will result in an optimal guide to assessing those patients for whom monitoring of amiodarone would be of absolute benefit.

1 High Performance Liquid Chromatographic Method for the Assay of Amiodarone and Desethylamiodarone in Plasma, Urine and Tissues. Pharmacokinetics of Amiodarone

Theo A. Plomp and Robert A. A. Maes

1.1 Introduction

Amiodarone is a benzofuran derivative with antianginal and antiarrhythmic properties. It is used widely for the treatment of ventricular and supraventricular arrhythmias, especially when resistent to conventional antiarrhythmic agents. In addition, it is considered one of the drugs of choice for management of tachycardias associated with the Wolff-Parkinson-White syndrome.

Amiodarone is well tolerated and does not usually cause serious untoward effects with maintenance doses of up to 600 mg per day. Adverse effects attributed to amiodarone includes photosensitivity, a bluish skin discoloration, formation of corneal microdeposits and disturbances of thyroid function, which appear to be dose and time related and reversible upon cessation of treatment.

From clinical practice it can be suspected that the drug is accumulated in the body and has a long "therapeutic" half-life since optimal antiarrhythmic effects may only be seen days to weeks after initiation of treatment and may persist for weeks after discontinuation of therapy. However, in contrast to its extensive clinical use, data on the pharmacokinetics and body distribution of the drug are still limited. Amiodarone is metabolised by N-deethylation and the desethyl metabolite is detectable in plasma of patients on long-term treatment. Data on the pharmacokinetic properties of desethylamiodarone, however, are scarce. In order to assess the above-mentioned pharmacokinetics and disposition of amiodarone and desethylamiodarone and to study the correlation between plasma levels and clinical effects, a fast, specific and sensitive method for the assay of amiodarone and desethylamiodarone in biological material has been developed.

1.2 Materials and Methods

The chromatographic instruments and conditions used for the assay are summarised in table 1.1. The HPLC system consisted of a Model 6000 A solvent delivery pump, a Model U6K injector, a guard column, a radial compression

Table 1.1: Chromatographic conditions.

Injector	Waters Assoc. U6K valve injector	
Pump	Waters Assoc. Model 6000 A	
Detector	Waters Assoc. Model 450 variable UV detector	
Column	10 cm × 8 mm I. D. Radial Pak C_{18} 10 μm	
Guard column	Bondapak C_{18}/Corasil, 37 – 50 μm	
Compression system	Model RCM-100 Module	
Mobile phase	methanol-ammonia 25% (99.3 : 0.7 v/v)	
Flow rate	2.0 ml/min	
Pressure	25 bar	
UV detection	242 nm	
Detector range	0.01 AUFS	
Injection volume	50 μl	
Retention times	desiodoamiodarone	4.0 min
	monoiodoamiodarone	5.2 min
	desethylamiodarone	5.3 min
	amiodarone	6.3 min
	internal standard	9.4 min

separation unit and a Model 450 variable wavelength detector, all from Waters Assoc. The radial compression separation unit consisted of a 10 cm × 8 mm I. D. Radial-Pak C18 (10 μm RP-18) cartridge and a Model RCM-100 Module for compressing of the cartridges. The output of the detector was displayed on a Varian Model A-25 recorder. The output signal also was fed to a Perkin Elmer LCI-100 data system for integration of peak areas. All chromatography was done at ambient temperature. The mobile phase consisted of methanol-ammonia 25% (99.3 : 0.7 v/v). The flow rate was 2.0 ml/min, maintained by a pressure of ca. 25 bars. The column effluent was monitored at 242 nm, the absorbance maximum of amiodarone, using a detector range of 0.01 absorbance unit full scale (AUFS) and a chart speed of 20 cm/h. Standard solution in methanol, plasma or urine containing amiodarone and desethylamiodarone at concentrations ranging from 0.10 – 5.0 μg/ml were made by appropriate dilution of the stock solution with methanol, plasma or urine respectively. An internal standard working solution of 2.5 μg/ml in acetonitrile was used. The extraction procedure for plasma, urine

Table 1.2: Procedure for the assay of amiodarone and desethylamiodarone in plasma and urine.

Plasma	Urine
1.00 ml of plasma	Add 2 ml of phosphate buffer, pH 7.0 to Clin Elut columns CE 1003
↓	↓
Add 2.00 ml of internal standard in acetonitrile	Add 40 μl of the internal standard solution and 1.00 ml of urine to the column
↓	↓
Vortex during 30 sec and let stand for 15 min	Add two 6 ml aliquots diisopropyl ether-acetonitrile (95 : 5), waiting 3 min after each addition
↓	↓
Centrifugate during 15 min at 2000 g	Collect combined eluants; evaporate. Reconstitute with 3.0 ml of methanol-acetonitrile (1 : 2)
↓	↓
Transfer supernatant to clean tube	Inject 50 μl into HPLC
↓	
Inject 50 μl into HPLC	

Table 1.3: Procedure for the assay of amiodarone and desethylamiodarone in tissues.

Tissue
Homogenise 100 mg of minced tissue with 1.00 ml of 50% ethanol
↓
Add 2.00 ml of internal standard in acetonitrile
↓
Vortex during 30 sec and let stand for 15 min
↓
Centrifugate during 15 min at 2000 g
↓
Transfer supernatant to clean tube
↓
Inject 50 μl into HPLC

and tissues is listed in tables 1.2 and 1.3. In a disposable polypropylene-capped glass centrifuge tube (16 × 100 mm) were introduced 1.00 ml of plasma and 2.00 ml of the solution of the internal standard in acetonitrile. The stoppered tube was vortexed for 30 sec, allowed to stand at room temperature for 15 min and centrifuged at 2000 g for 15 min. Subsequently the supernatant was transferred to a clean, disposable, glass tube and 50 μl of this solution were injected into the liquid chromatograph. Using the direct deproteinisation method for urine specimens, even after injection of 10 μl of the supernatant, a very broad solvent peak was observed caused by endogenous UV absorbing substances from urine. For this reason a column clean-up procedure was used. To a Clin-Elut 1003 disposable extraction column was succesively added 2.00 ml phosphate buffer pH 7, 40 μl of the stock internal standard solution and 1.00 ml urine. After waiting for 3 min, 6 ml of a mixture of diisopropyl ether-acetonitrile (95 : 5) was added to the column. The eluate was collected in a 25 ml conical flask and after at least 3 min the extraction was repeated with one more 6 ml aliquot. After completion of the elution from the column, the collected combined eluates were evaporated to dryness under reduced pressure in a Buchi Rotavapor at room temperature. The sample residue then was reconstituted with 3.0 ml of a mixture of methanol-acetonitrile (1 : 2 v/v) and 50 μl aliquots were injected into the chromatograph. Tissue was finely minced, dried between Kleenex® tissues, and portions of about 100 mg of the dried minced tissue were homogenised during 10 min with 1.00 ml of 50% ethanol in a glass tube with a Potter homogeniser. The tissue homogenate then was vortexed for 15 min with 2.0 ml of the internal standard solution and thereafter completely transferred to a clean glass test tube and centrifuged at 2000 g for 15 min. From the supernatant 50 μl were injected in the liquid chromatograph. All samples were assayed in duplicate. The concentration of amiodarone and desethylamiodarone in plasma, urine and tissues was determined from calibration curves of peak area ratios (amiodarone and desethylamiodarone to internal standard) versus amiodarone and desethylamiodarone concentrations in plasma, urine and tissue standards carried through the described procedures.

1.3 Results

Chromatograms from blank human plasma (a), urine (c), tissue (e) and those spiked with known concentrations of amiodarone and desethylamiodarone (b. d. f.) are shown in fig 1.1. Chromatograms of a plasma sample (a), and a renal (b) and heart (c) tissue sample obtained respectively from a patient treated with 200 mg amiodarone per day for more than a year and from post-mortem case treated with the same dosage for more than one year are presented in fig 1.2. The peaks representing amiodarone, desethylamiodarone and the internal standard are symmetrical and well removed from the solvent front and interfering peaks from the biological material. The retention times of desethylamiodarone, amiodarone and internal standard were 5.3, 6.3 and 9.4 min respectively. For the

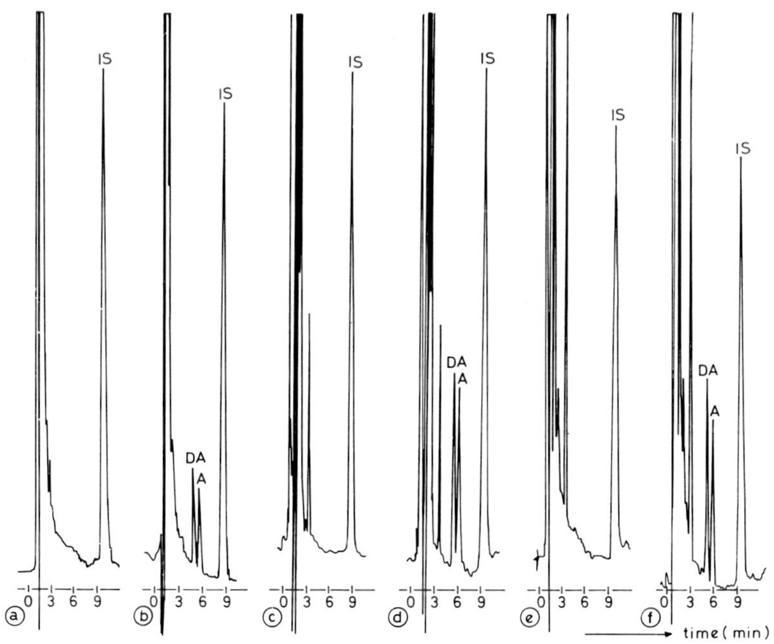

Fig 1.1: Chromatograms of (a) blank human plasma, (b) plasma spiked with 1.0 μg/ml amiodarone (A) and desethylamiodarone (DA), (c) blank urine, (d) urine spiked with 2.5 μg/ml A und DA, (e) blank heart tissue, and (f) heart tissue spiked with 25 μg/g A and DA. Internal standard (IS) concentration 3.3 μg/ml. Injection volume: 50 μl. Detector sensitivity: 0.01 a.u.f.s. Recorder chart speed: 20 cm/h.

potential metabolites (desiodo, L3937) and (monoiodo, L6355) retention times of 4.0 and 5.2 min were observed respectively.

A number of drugs were investigated for possible interference with the amiodarone assay. Verapamil, cinnarizine, quinidine und disopyramide were detectable if present at concentrations of about 1 – 5 μg/ml and showed in this system retention times of 2.5, 3.2, 4.7 and 4.8 min respectively. However, they were all completely resolved from the compounds of interest on the chromatographic system. No interfering peaks were observed in the plasma of patients receiving amiodarone in combination with a variety of drugs belonging to the groups of antihypertensive agents, diuretics, vasodilators, β-blocking agents, cardiac glycosides, antiarrhythmic agents, tranquillizers, analgesics and anticoagulants.

The calibration curves in plasma were established every week, and standard plasma samples were analysed during the week to validate the method. The equa-

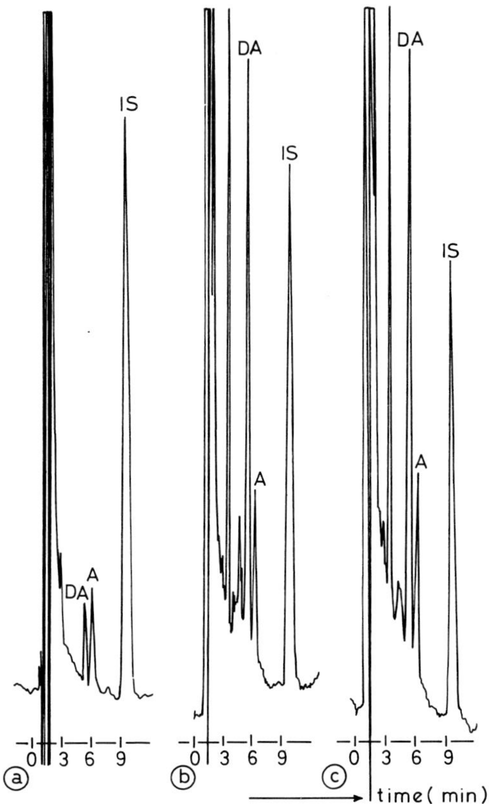

Fig 1.2: Chromatograms of (a) plasma of a patient receiving 200 mg amiodarone for more than a year, containing 1.03 µg/ml A and 0.70 µg/ml DA, (b) kidney tissue containing 20.5 µg/g A and 57.4 µg/g DA, and (c) heart tissue containing 22.0 µg/g A and 56.1 µg/g DA from a post-mortem case who had received 200 mg of amiodarone for more than a year. Conditions as in fig 1.1.

tion of the curves was calculated by least-squares linear regression. For the curves in plasma and urine a good linear relationship was obtained in the concentration range studied for both drugs with y intercepts not significantly different from zero. The linear regression data for the mean calibration curves of amiodarone and desethylamiodarone in plasma and urine are summarised in table 1.4.

The plasma and urine curves were the average of 8 respectively 4 standard curves, which were run during a two respectively one month time period. The precision of the assay was determined by replicate analysis of spiked plasma

Table 1.4: The linear regression parameters for the mean calibration curves of amiodarone and desethylamiodarone in plasma and urine.

Compound	Sample	n*	Linear regression parameters**		
			slope	y-intercept	correlation coefficient
Amiodarone	plasma	8	0.1091	0.0032	0.9998
	urine	4	0.1237	−0.0008	0.9979
Desethylamiodarone	plasma	8	0.1100	0.0027	0.9999
	urine	4	0.1161	0.0000	0.9992

* Number of calibration curves used

** Peak area ratio of drug to internal standard plotted on the y-axis versus drug concentration in plasma or urine (in μg/ml) on the x-axis

samples containing amiodarone and desethylamiodarone at concentrations of 0.75, 2 and 4 μg/ml. The intra-assay and inter-assay coefficients of variation are presented in table 1.5.

Table 1.5: Precision data for the determination of amiodarone and desethylamiodarone in plasma.

Concentration μg/ml	Coefficient of variation (%) Within-day (n = 7)		Day-to-day* (n = 6)	
	Amiodarone	Desethyl-amiodarone	Amiodarone	Desethyl-amiodarone
0.75	3.1	4.6	5.6	5.6
2.00	1.5	3.4	3.5	3.7
4.00	1.2	1.4	2.2	2.8

* Analysis performed on 6 days during a 3 week period

The within-day and day-to-day variation at each concentration for both drugs was less than 6%. The sensitivity of the assay was 25 ng/ml for both drugs using an injection volume of 100 μl of plasma sample extract. The within-day coefficient of variation for spiked plasma samples containing 25 ng/ml was ±17% (n = 5).

The analytical recovery of amiodarone and desethylamiodarone from plasma, urine and various tissues was determined by comparison of the peak areas of amiodarone and desethylamiodarone obtained by analysis of 50 μl portions of methanolic standards mixed with internal standard solution (1 : 2 v/v) to those obtained from freshly prepared sample extracts. The results of the recovery studies are presented in tables 1.6 and 1.7. The recovery of the internal standard

Table 1.6: Recovery of amiodarone (A) and desethylamiodarone (DA) from plasma and urine.

Concentration (μg/ml)	Recovery %			
	Plasma (n = 8*)		Urine (n = 5)	
	A	DA	A	DA
0.10	99.7	104.0	84.2	76.2
0.25	102.3	105.6	85.3	86.4
0.50	100.7	105.5	84.4	74.8
1.00	104.8	106.7	90.2	85.2
1.75	102.2	104.3	90.9	77.7
2.50	104.2	100.8	81.0	72.4
3.50	103.5	101.5	89.2	81.6
5.00	98.9	102.0	90.5	75.6
Mean	102.0	103.8	87.0	78.7
SD	2.1	2.2	3.7	5.1

* Number of determinations at each concentration

Table 1.7: Recovery of amiodarone (A) and desethylamiodarone (DA) from tissues.

Concentration (μg/g)	Recovery % Tissues** (n = 12*)	
	A	DA
1.00	86.1	89.3
2.50	97.0	94.7
5.00	99.6	98.2
Mean	94.2	94.1
SD	7.2	4.5

* Number of determinations at each concentration
** Mean recovery data from aortic arch, oesophagus, lung, heart, liver and renal tissue

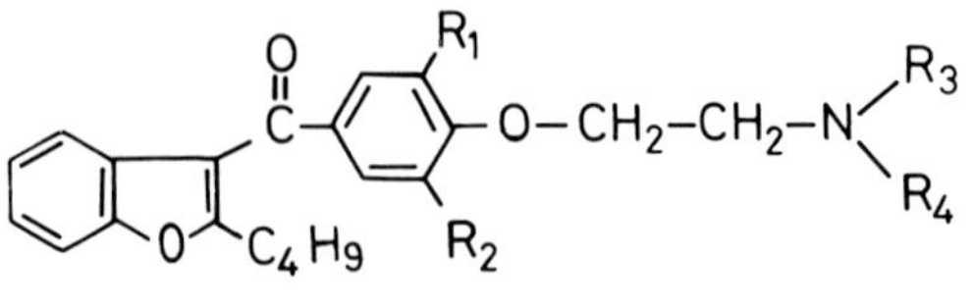

Compound	R_1	R_2	R_3	R_4
Amiodarone L3428	I	I	C_2H_5	C_2H_5
Desethylamiodarone L 32812	I	I	C_2H_5	H
L 6355	I	H	C_2H_5	C_2H_5
L 3937	H	H	C_2H_5	C_2H_5

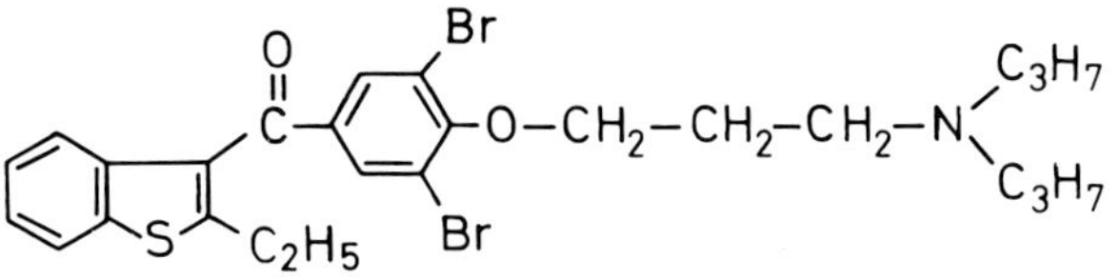

Internal standard, L 8040

Fig 1.3: Chemical structures of (top) amiodarone, desethylamiodarone, two potential metabolites, and (bottom) internal standard.

Table 1.8: Modified chromatographic conditions.

Injector	Waters Assoc. U6K valve injector	
Pump	Waters Assoc. Model 6000 A	
Detector	Waters Assoc. Model 481 Labda Max variable UV detector	
Column	10 cm × 8 mm I.D. Nova Pak C_{18}, 5 µm	
Precolumn	Guard Pak Precolumn module with C_{18} inserts	
Compression system	Model RCM-100 module	
Mobile phase	methanol-ammonia 25% (99.3 : 0.7 v/v)	
Flow rate	1.0 ml/min	
Pressure	25 bar	
UV detection	242 nm	
Detector range	0.005 AUFS	
Injection volume	20 µl	
Retention times	desethylamiodarone	6.3 min
	amiodarone	7.9 min
	internal standard	10.8 min

from plasma and urine was 107% (n = 70) and 76% (n = 40) respectively. For the 2 potential metabolites L3937 and L6355 (fig 1.3) recoveries in plasma of 103.3% (n = 4) and 103.8% (n = 4) respectively were observed at a concentration of 1.0 µg/ml. In urine at a concentration of 1 µg/ml recoveries of 85.5% (n = 4) for L3937 and 66.3% (n = 4) for L6355 were found. The modifications of our HPLC method introduced in 1985 are summarised in table 1.8 [1]. Typical chromatograms using these modified chromatographic conditions are depicted in figs 1.4 and 1.5.

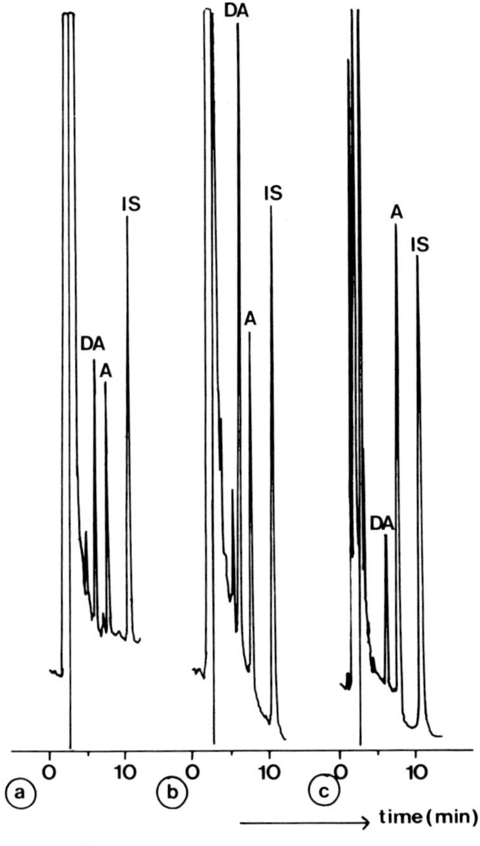

Fig 1.4: Chromatogram of (a) plasma containing 1.41 µg/ml A and 1.14 µg/ml DA, (b) heart tissue containing 19.1 µg/g A and 56.6 µg/g DA, and (c) adipose tissue containing 226 µg/g A and 60 µg/g DA from a heart-transplantation patient who had received 300 mg of amiodarone for two years.

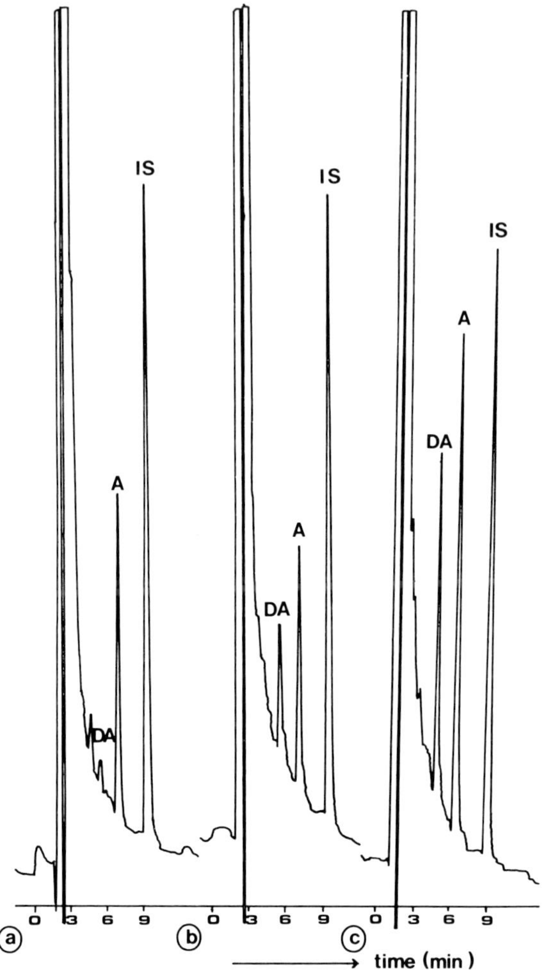

Fig 1.5: Chromatogram of (a) serum sample containing 1.35 µg/ml A and 0.20 µg/ml DA, (b) heart tissue sample containing 12.4 µg/g A and 5.8 µg/g DA, and (c) renal tissue sample containing 70.7 µg/g A and 45.0 µg/g DA obtained from a rat orally treated with 100 mg/kg of amiodarone per day for 13 days.

Fig 1.4 shows the chromatograms of a plasma sample (a) and a heart (b) and adipose (c) tissue sample obtained from a patient who underwent a heart transplantation and was treated with 300 mg amiodarone per day for more than two years.

Fig 1.5 shows the chromatograms of an extract of a serum sample (a) and a renal (b) and heart (c) tissue sample obtained from a rat orally treated with 100 mg/kg of amiodarone per day for 13 days.

The described method was used to determine the plasma levels of amiodarone and desethylamiodarone after single oral and intravenous administration and after repeated oral administration of the drug. The plasma concentration time curves of amiodarone obtained after single intravenous and oral administration of 400 mg to 7 healthy volunteers are shown in figs 1.6 and 1.7. The curves were fit by using a model independent curve fitting program (Pharmfit) developed by Van Rossum and co-workers [2].

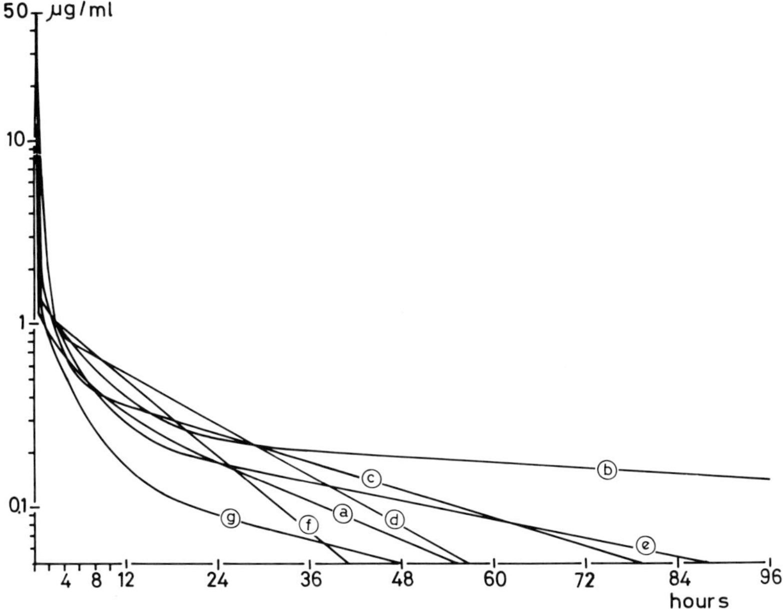

Fig 1.6: The individual amiodarone plasma concentration-time curves following single intravenous administration of 400 mg of amiodarone to healthy subjects a – g. Solid lines are the computer fits to biexponential (f) and triexponential (a – e, g) equations.

The mean amiodarone plasma levels after single oral and intravenous administration of the drug to the 7 healthy volunteers are outlined in table 1.9.

After intravenous administration, plasma levels were about 11 µg/ml after 15 min and then decreased to 1.7 µg/ml after 1 h and to 1 µg/ml after 2 h. After oral

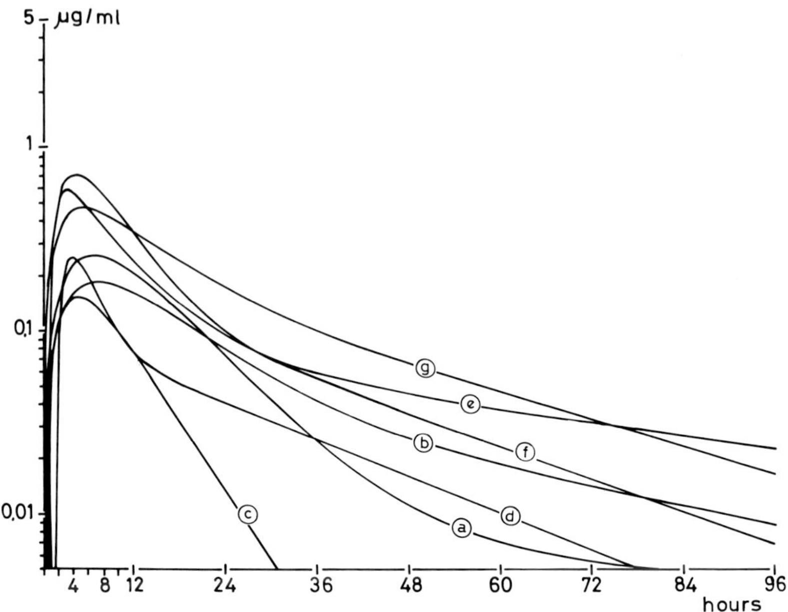

Fig 1.7: The individual amiodarone plasma concentration-time curves following single oral administration of 400 mg of amiodarone to healthy subjects a – g. Solid lines are the computer fits to biexponential (c) and triexponential (a, b, d – g) equations.

administration the absorption is slow and variable and plasma levels at 1 h are about 0.1 µg/ml and reached a maximum of 0.4 µg/ml after 5 h.

The pharmacokinetic parameters calculated for the 7 individuals after a single intravenous and oral administration are shown in tables 1.10 and 1.11. The mean steady-state plasma levels of amiodarone and desethylamiodarone with various daily maintenance dosages are shown in table 1.12.

The elimination half-lives (t 1/2) calculated from the plasma concentration time curves of 4 patients after cessation of treatment varied for amiodarone from 27 – 52 days and for desethylamiodarone from 30 – 94 days (table 1.13). The drugs were still detectable in plasma ±200 days after discontinuation of treatment.

Neither amiodarone nor desethylamiodarone were detected in urine samples of several patients on long-term amiodarone treatment. Very high levels of amiodarone and desethylamiodarone were found respectively in lung and liver tissue of two post-mortem cases and in adipose tissue of two surgical patients (tables 1.14, 1.15).

Table 1.9: Mean amiodarone plasma concentrations after a single intravenous and oral dose of 400 mg of amiodarone to 7 healthy volunteers.

Time after administration in hours	Amiodarone plasma concentration (μg/ml)	
	Oral study	Intravenous study
0.125	0.0	16.2 ± 6.4
0.25	0.004 ± 0.008	10.8 ± 12.2
0.50	0.037 ± 0.045	3.20 ± 2.88
1.0	0.116 ± 0.131	1.69 ± 1.04
2.0	0.270 ± 0.184	1.04 ± 0.24
4.0	0.326 ± 0.222	0.81 ± 0.12
6.0	0.371 ± 0.203	0.64 ± 0.17
8.0	0.267 ± 0.169	0.46 ± 0.13
12.0	0.210 ± 0.050	0.38 ± 0.12
24.0	0.076 ± 0.050	0.20 ± 0.06
48.0	0.041 ± 0.039	0.09 ± 0.07
96.0	0.006 ± 0.008	0.03 ± 0.05
Mean body weight:	66 ± 8 kg	
Mean age:	28 ± 5 yr	

Table 1.10: Pharmacokinetic parameters of amiodarone following the intravenous administration of 400 mg to 7 subjects.

Parameter	Range	Mean ± S. D.
Age (yr)	20 – 35	28 ± 5
Body weight (kg)	53 – 78	66 ± 8
MRT (h)	6 – 143	33 ± 49
t 1/2 (β) (h)	9 – 116	35 ± 37
$AUC_{0-\infty}$ (mg h/l)	19 – 58	36 ± 18
Cl (l/h)	7 – 22	14 ± 7
Vf (l)	43 – 1150	359 ± 381

The data on amiodarone plasma concentration after single and repeated oral administration are in good agreement with the levels observed by Andreasen et al. [3].

In addition the very large volume of distribution, long t 1/2 (β), the tmax and Cmax and the slow and variable absorption are in good agreement with the findings of Riva [4], Haffajee [5], Kannan [6] and Andreasen and co-workers [3].

Table 1.11: Pharmacokinetic parameters of amiodarone following the oral administration of 400 mg to 7 subjects.

Parameter	Range	Mean ± S. D.
Age (yr)	20 – 35	28 ± 5
Body weight (kg)	53 – 78	66 ± 8
t 1/2 (β) (h)	9 – 116	35 ± 37
t_{max} (h)	2.7 – 6.9	4.8 ± 1.5
C_{max} (mg/l)	0.15 – 0.70	0.4 ± 0.2
$AUC_{0-\infty}$ (mg h/l)	3 – 13	8 ± 4
F (%)	10 – 71	30 ± 23

Table 1.12: Plasma amiodarone (A) and desethylamiodarone (DA) levels in chronic oral therapy.

Maintenance dose per day (mg)	n	Mean plasma level* (µg/ml)		R DA/A
		A	DA	
200	32	0.99 ± 0.40	0.85 ± 0.31	0.86
300	12	1.47 ± 0.57	1.29 ± 0.54	0.88
400	30	1.50 ± 0.61	1.25 ± 0.59	0.83
600	13	2.90 ± 1.00	2.10 ± 0.70	0.72
800	15	3.34 ± 1.22	1.76 ± 0.61	0.53

n Number of patients
* Mean ± s.d. of the mean steady level of each patient

Table 1.13: Elimination kinetics of amiodarone and desethylamiodarone after long-term oral amiodarone therapy.

Term	Units	Patients			
		1	2	3	4
Age/Sex	yr	56/M	29/F	43/M	25/F
Weight	kg	71	55	95	71
Daily dose	mg	600	400	400	600
Duration of therapy	month	2	9	24	48
Conc. amiodarone	µg/ml	1.00	1.86	1.07	3.09
Conc. desethylamiodarone	µg/ml	1.32	2.30	0.96	2.73
t 1/2 amiodarone	day	26.65	40.71	39.62	51.73
t 1/2 desethylamiodarone	day	29.70	54.52	51.57	93.56

Table 1.14: Post-mortem tissue concentration of amiodarone (A) and desethylamiodarone (DA) of 2 cases on long-term amiodarone treatment.

	Concentration (μg/g or μg/ml)			
	Case 1		Case 2	
	A	DA	A	DA
Lung	178	541	28	238
Liver	112	307	7	64
Kidney	16	53	3	19
Heart	18	52	4	19
Thyroid	–	–	3	13
Aorta	10	13	–	–
Esophagus	8	21	–	–
Plasma	0.8	0.4	–	–

Treatment case 1: 200 mg amiodarone per day for about 1 year
Treatment case 2: 200 mg amiodarone per day during 7 weeks

Table 1.15: Tissue concentrations of amiodarone (A) and desethylamiodarone (DA) in 2 surgical patients on long-term amiodarone treatment.

	Concentration (μg/g or μg/ml)			
	Patient 1		Patient 2	
	A	DA	A	DA
Adipose tissue	364	135	112	65
Myocardium	49	114	12	64
Endocardium	29	84	–	–
Plasma	1.5	1.7	0.8	0.9

Treatment patient 1: 600 mg amiodarone per day for 8 months
Treatment patient 2: 200 mg amiodarone per day for 2 months

Finally the presented method has shown to be a valuable tool for the monitoring of amiodarone and its major metabolite in steady state patients and for the further elucidation of amiodarone disposition and pharmacokinetics in men.

1.4 References

[1] Plomp T. A., Wiersinga W. M., Maes R. A. A.: Tissue distribution of amiodarone and desethylamiodarone in rats after repeated oral administration of various amiodarone dosages. Arzneim. Forsch. Drug Res. 35 (1985) 1805–1810.

[2] Plomp T. A., Van Rossum J. M., Robles de Medina E. O., Van Lier T., Maes R. A. A.: Pharmacokinetics and body distribution of amiodarone in man. Arzneim. Forsch. Drug Res. 34 (1984) 513–520.

[3] Andreasen F., Agerbeek H., Bjerregaard P., Gotzske H.: Pharmacokinetics of amiodarone after intravenous and oral administration. Eur. J. Clin. Pharmacol. 19 (1981) 293–299.

[4] Riva E., Gerna M., Latini R., Diani P., Volpi A., Maggioni A.: Pharmacokinetics of amiodarone in man. J. Cardiovasc. Pharmacol. 4 (1982) 264–269.

[5] Haffajee C. I., Love J. C., Canada A. T., Lesko I. J., Asdourian G., Alpert J. S.: Clinical pharmacokinetics and efficacy of amiodarone for refractory tachyarrhythmias. Circulation 67 (1983) 1347–1355.

[6] Kannan R., Nademanee K., Hendrickson J. A., Rostani H. J., Singh B. N.: Amiodarone kinetics after oral doses. Clin. Pharmacol. Ther. 31 (1982) 438–444.

2 Therapeutic Drug Monitoring of Amiodarone

David W. Holt

Amiodarone is a Class III antiarrhythmic drug which has proved remarkably effective for the treatment of a wide spectrum of arrhythmias refractory to other therapy. There are several factors which make the routine measurement of amiodarone as a guide to therapy a clinically useful objective.

Firstly, the drug has complex, unusual, pharmacokinetics. Secondly, the compound is often prescribed for prolonged periods, perhaps, for life. Thirdly, the compound has been implicated in the development of serious, life-threatening, adverse effects. Finally, compliance is an important factor to be established in the assessment of efficacy. It is also worth noting that both the parent compound and its principal metabolite are comparatively easy to measure.

This review will cover three main areas with which we have been involved during the past decade. They are the impact of the pharmacokinetics of amiodarone on the interpretation of its measurement, the determination of rational dosage regimes and the use of amiodarone measurements for the avoidance of toxicity.

Our current methodology is summarised in table 2.1 from which it can be seen that a very simple HPLC technique is capable of producing results for the routine measurement of this compound [1]. Modifications of this method have been developed for the measurement of amiodarone following single oral dosage and in tissues from amiodarone-treated patients [2, 3].

2.1 Pharmacokinetics: Basic Parameters and Dosing

The pharmacokinetics of amiodarone are unusual and incompletely characterised. Broekhuysen and co-workers [4] noted a mean terminal elimination half-life of 28 days for amiodarone but this finding was based on the measurement of radioactively labelled amiodarone. It was some years later that methodology with both the selectivity and sensitivity to investigate the single-dose pharmacokinetics of amiodarone was published.

Several centres have derived pharmacokinetic parameters for amiodarone following single-dose intravenous and oral administration of the drug, but there is considerable variation in their findings [5 – 12]. This is largely due to differences in the time period over which blood samples were collected and the sen-

Table 2.1: HPLC analysis of amiodarone.

Extraction	
Sample/standard	200 µl
Buffer/internal standard	50 µl
Methyl t-butyl ether	200 µl
Vortex mix	
Centrifuge (10000.g lmin)	
Inject	100 µl
Chromatography	
Mobile phase	10 mM ammonium perchlorate
Column	Spherisorb 5 SW
Detector	UV, 240 nm

sitivities of the assays used to measure the drug. Table 2.2 summarises the data from several studies, from which it is apparent that in those studies in which sampling continued for more than a month much larger values were derived for the terminal elimination half-life and volume of distribution. There is less discrepancy in the measurement of oral bioavailability, for which it has been established that sampling over a short time period is valid [13].

Table 2.2: Mean pharmacokinetic parameters in relation to duration of study.

Reference	Cl (l/h)	V (l/Kg)	"Terminal" half-life	Bioavailability (%)	Sampling period
Andreasen et al. [4]	–	–	–	43	96 h
Anastasiou-Nana et al. [5]	10.7	1.0	4.4 h (iv) 17 h (po)	35	30 h
Riva et al. [6]	35.7	12.9	17.4 h (iv) 36 h (po)	58	50 h
Plomp et al. [8]	14.7	5.7	33 h	31	96 h
Pourbaix et al. [10]	–	–	–	65	72 h
Berger et al. [9]	9.4	106	18 d	–	42 d
Holt et al. [7]	8.6	62	25 d	35	56 d
Nolan et al. [11]	11.6	173	30 d	38	90 d

Cl = total body clearance, V = volume of distribution.

The data from these studies indicate that amiodarone has poor oral bioavailability, a low total body clearance, a very large volume of distribution and an exceptionally long terminal elimination half-life. The reason for the poor oral bioavailability of such a lipid soluble drug has not been established but does not seem to be associated with either presystemic hepatic clearance [8, 12] or poor dissolution characteristics of the tablet preparation [11].

Only one metabolite of amiodarone has been identified in man, namely desethylamiodarone [14]. The pharmacological effects of this compound have not been defined. Deiodination of amiodarone has been proposed and in the dog di-N-desethylamiodarone has been identified [15].

In patients receiving chronic amiodarone therapy in the dose range 100–600 mg/day there is a good correlation between dose and plasma amiodarone concentration; mean plasma concentrations following daily doses of 200 and 400 mg/day are approximately 1 and 2 mg/l, respectively [8].

In patients, the terminal elimination half-life of amiodarone following cessation of chronic therapy is generally longer than that noted after single dosage, averaging about 50 days [5, 8, 9]. This slower elimination may be due, in part, to inhibition of hepatic drug metabolising enzymes by amiodarone; an increase in the half-life of antipyrine following amiodarone therapy of six weeks duration has been noted [16]. After amiodarone dosage the half-life of the desethyl metabolite in patients is similar, but longer than that of the parent compound, averaging about 60 days [8, 9].

Negligible quantities of both compounds are excreted unchanged in urine [17] and their elimination from plasma is unaffected by haemodialysis [18]. The effects of age, cardiac failure and hepatic failure on the pharmacokinetics of amiodarone have not been described.

Because the drug has a very long elimination half-life, plasma concentrations rise slowly. Recently, a study was undertaken with the Royal Infirmary, Lancaster and Hammersmith Hospital, London, to examine two questions. 1. Can an i. v. dosage schedule be used easily, to achieve concentrations consistent with efficacy rapidly? 2. Are adverse effects due to a slow accumulation of either amiodarone or the desethyl metabolite during chronic therapy? Two groups were studied:

Group 1. 84 patients (68 M/16 F) mean (SD) age 62(8)y all with suspected acute myocardial infarction. All patients received i. v. amiodarone prophylaxis as follows: 5 mg/Kg for 30 min, then 1000 mg in 24h, thereafter, 600 mg/d p. o. for 1 week, then 200 mg/d. Frequent samples for drug measurement were collected during the period 0 – 48 hours and then for up to 10 days.

Group 2. 49 patients (29 M/20 F) mean age 56(16)y. Patients were receiving chronic amiodarone therapy for the treatment of both atrial and ventricular arrhythmias. They had received amiodarone 100–600 mg/d for at least 3 months. All the patients were monitored for at least 11 months without any alteration of their maintenance dose.

The results for Group 1, following the i. v. dose of amiodarone, are shown in table 2.3. Mean amiodarone concentrations after 2 hour infusion were well within a target range of 1 – 2 mg/l. Desethylamiodarone concentrations increased very slowly. After 5 – 10 days of oral therapy mean concentrations of amiodarone and desethylamiodarone were 1.2 and 0.7 mg/l, respectively; the mean ratio amiodarone/desethylamiodarone was 1.7, markedly higher than the ratio which is seen during chronic therapy.

Table 2.3: Amiodarone and desethylamiodarone concentrations Group 1.

Group 1: i. v. amiodarone

Mean (SD) plasma amiodarone concentrations (mg/l):

Time (h)						
0.5	2	4	8	12	24	48
6.6(4.0)	1.8(1.0)	1.5(0.6)	1.5(0.6)	1.6(0.7)	1.7(0.7)	1.0(0.5)

Desethylamiodarone*

Mean concentrations:	24 h	0.1 mg/l
	48 h	0.2 mg/l

* Not detected in 8/84 patients at 24 h (Limit of detection 0.25 mg/l).

The patients in Group 2 were monitored for 11 – 67 months; the mean period of follow-up was 19.6 (13.4) months and the total number of observations was 210.

A linear regression analysis was performed, relating amiodarone and desethylamiodarone plasma concentrations with months of therapy, for each patient. The observed slopes range from −0.024 to +0.16 (median slope −0.001 mg/1/m). This slope was not significantly different from zero.

The mean ratio, amiodarone/desethylamiodarone, was 1.05(0.3) and this ratio did not vary with duration of therapy ($r=0.04$, $p>0.7$).

From these two related studies it was concluded that a simple i. v. dosage regimen could achieve plasma amiodarone concentrations consistent with efficacy rapidly, with minimal adverse effects. Metabolite concentrations rise slowly. Despite their very long terminal elimination half-life, there was no evidence of prolonged accumulation of either compound in plasma. During chronic therapy the ratio amiodarone/desethylamiodarone does not vary, and this ratio is a good indication of duration of therapy.

2.2 Tissue Distribution

A very large volume of distribution is consistent with tissue accumulation of amiodarone. In biopsy and autopsy material from patients treated with the drug the highest concentrations were found in subcutaneous fat, liver and lung, whilst the lowest concentrations were found in brain [19].

In all tissues studied, with the exception of fat, the concentrations of desethylamiodarone were higher than the parent compound. Typical values for the tissue concentrations of both compounds are shown in table 2.4. They represent significant body stores of amiodarone, of which an estimated 40% is present in fat; a typical body load could be equivalent to as much as 30 days therapy. These findings suggest that obesity could have an important influence on both dosage requirements and the persistence of clinical effect observed following cessation of amiodarone therapy [20].

Table 2.4: Tissue concentrations of amiodarone and desethylamiodarone.

Tissue	Amiodarone	Desethylamiodarone
Fat	316	76
Liver	391	2354
Lung	198	952
Adrenal	137	437
Testis	89	470
Kidney	57	262
Lymph node	83	316
Heart	40	169
Skeletal muscle	22	51
Thyroid	14	64
Brain	8	54

Concentrations – mg/Kg wet weight; based on Ref. [19]

Amiodarone is highly bound to plasma proteins, principally albumin, with only 4% of the drug remaining unbound [21], a finding which is consistent with measurements of amiodarone in saliva [22]. Plasma amiodarone concentrations are much lower in fetal compared with maternal circulation, [23–25] but comparatively high amiodarone concentrations have been found in breast milk in one case [23].

The widespread tissue deposition of amiodarone and the desethyl metabolite is associated with morphological changes, characterised by the appearance of multilamellar lysosomal inclusion bodies, which are most clearly seen in tissues containing the highest concentrations of both compounds. Interestingly, these tissues are those associated with the more serious adverse-effects of the drug, notably the lung, liver and pigmented skin [19, 26–28]. These observations led us to conduct a study in association with the Freeman Hospital, Newcastle-upon-Tyne, to assess the relationship between plasma amiodarone concentrations and morphological changes in peripheral neutrophils.

We had demonstrated the frequent development of multilamellar inclusion bodies in various tissues, including neutrophils, suggesting a drug-induced lipidosis [19]. This appearance was seen not only in association with clinical adverse effects but also in their absence. We investigated the frequency with which multilamellar bodies appeared in the neutrophils of peripheral blood samples from patients taking long-term amiodarone. Their presence or absence was correlated with drug and metabolite concentrations in plasma and putative lipidosis-associated adverse effects, in particular those affecting the liver and lung.

14 patients (9 M/5 F), mean (SD) age 59.7(8.8)y (range 41–70), who had received a mean amiodarone dose of 403(136)mg/d (range 200–600) for 16.4 (12.2) m (range 2.5–41.5) were studied. Patients were receiving amiodarone for the treatment of cardiac arrhythmias. The patients were assessed for the development of possible hepatic (plasma AST) and pulmonary adverse effects (pulmonary function testing included carbon monoxide transfer factor). Neuropathy was noted clinically but cutaneous and corneal changes were not assessed routinely.

Separate EDTA anticoagulated samples were collected for the preparation of buffy coat specimens which were examined by transmission electron microscopy at a magnification of $\times 10000$ after routine heavy metal staining. Randomly selected fields were photographed and the prints were assigned to one of two groups by an independent observer, blind to patient details. In Group 1 multilamellar bodies were absent or equivocal whilst in Group 2 there were at least 3 multilamellar bodies per neutrophil.

There were seven patients in each group. In Group 1 only one patient had unwanted effects as defined, a raised AST (64IU/l), and his plasma amiodarone and desethylamiodarone concentrations were high (4.4/2.3 mg/l); one patient was hypothyroid. In Group 2 only one patient had no unwanted effects but four had AST at least twice normal, and three had frank pulmonary toxicity; one patient suffered peripheral neuropathy, whilst another had a sustained fall in carbon monoxide transfer factor of 26% and was hypothyroid.

Mean plasma concentrations of amiodarone and desethylamiodarone (Am/DAm) were higher in Group 2 (2.7(0.5) / 2.9(1.1) mg/l) than in Group 1 (1.5(1.4) / 1.2(0.6)mg/l), $p < 0.02$ (Mann-Whitney U test). There was a significant association between the appearance of multilamellar bodies and plasma amiodarone concentrations in excess of 2 mg/l, $p < 0.04$ (X^2 test with Yates correction). Although the mean daily dose of amiodarone was higher in Group 2 (363(171)mg vs 443(83)mg), there was no significant difference between the two groups in either the duration of therapy or the total dose.

It was concluded that, in this preliminary study of a patient population with a high incidence of amiodarone toxicity, the presence of multilamellar bodies in blood neutrophils was related to a number of clinically important unwanted effects, suggesting a systemic lipidosis. High amiodarone and desethylamiodarone concentrations were associated with the detection of multilamellar bodies, suggesting that the development of a lipidosis may be dose related. Electron microscopy of peripheral blood neutrophils, as well as the measurement of plasma drug concentrations, could be of value in assessing the risk of toxicity in patients receiving the drug chronically.

2.3 Drug Interactions

Interactions between amiodarone and a number of drugs have been reviewed [29]. Two interactions with a documented pharmacokinetic basis are of particular clinical importance.

Amiodarone potentiates the anticoagulant effects of warfarin [30], the reduction in warfarin dosage necessary when the drugs are co-administered being related to the daily dose of amiodarone. The underlying mechanism for this interaction is an inhibition of warfarin elimination; the plasma protein binding of warfarin is unaffected by amiodarone [31].

In adult patients plasma concentrations of digoxin rise by 70–100% when amiodarone is prescribed concomitantly [32, 33] and this interaction has been associated with digoxin toxicity [33, 34]. A decrease in both renal and non-renal clearance of digoxin has been proposed to explain the interaction [33–35], inhibition of renal tubular secretion being particularly important in paediatric patients [34].

Recently, we have shown that the increase in plasma concentrations of digoxin brought about by amiodarone is sustained during chronic amiodarone therapy and is not only related to the loading dose period [36].

2.4 Conclusions

The pharmacokinetics of amiodarone differ markedly from those of the other antiarrhythmic drugs in current use. The data presented here give some guidance for the optimal administration of the compound.

If a rapid onset of clinical effect is needed a loading dose must be used since the drug has a relatively poor oral bioavailability. Large oral loading doses may be associated with both gastrointestinal and neurological side effects [37] and the use of an intravenous loading dose has been advocated [38].

There is extensive tissue accumulation of amiodarone and its desethyl metabolite. The association between the highest tissue concentrations of these compounds and unwanted drug effects underlines the advice that the minimum effective dose of amiodarone should be used [37]. Dose titration is, of course, complicated by the exceptionally long elimination half-life of the drug and clinicians must be prepared to make prolonged observations after making a dosage adjustment.

The measurement of amiodarone and desethylamiodarone concentrations in plasma may be helpful as a guide to compliance and dosage requirements, especially in the avoidance of adverse effects [26, 39–41]. The data do support the conclusion that plasma concentrations are related to the administered dose, morphological changes in some tissues and serious toxic effects. The relationship with efficacy is less clear but, to avoid the most serious complications of amiodarone therapy, a reference range of 1–2 mg/l should be advised.

2.5 References

[1] Flanagan R. J., Storey G. C. A ., Holt D. W.: Rapid high-performance liquid chromatographic method for the measurement of amiodarone in blood-plasma or serum at the concentrations attained during therapy. J. Chromatogr. 187 (1980) 391–398.

[2] Storey G. C. A., Holt D. W.: High-performance liquid chromatographic measurement of amiodarone and desethylamiodarone in plasma or serum at the concentrations attained following a single 400 mg dose. J. Chromatogr. 245 (1982) 377–380.

[3] Storey G. C. A ., Adams P. C., Campbell R. W. F., Holt D. W.: High-performance liquid chromatographic measurement of amiodarone and desethylamiodarone in small tissue samples following enzymatic digestion. J. Clin. Pathol. 36 (1983) 785–790.

[4] Broekhuysen J., Lauruel R., Sion R.: Recherches dans la serie des benzofuranes. XXXVII. Etude comparée du transit et du metabolisme de l'amiodarone chez diverses espèces animales et chez l'homme. Arch. Int. Pharmacodyn. 177 (1969) 340–359.

[5] Andreasen F., Agerbaek H., Bjerregaard P., Gotzsche H.: Pharmacokinetics of amiodarone after intravenous and oral administration. Eur. J. Clin. Pharmacol. 19 (1981) 293–299.

[6] Anastasiou-Nana M., Levis G. M., Moulopoulos S.: Pharmacokinetics of amiodarone after intravenous and oral administration. Int. J. Clin. Pharmacol. Ther. Tox. 20 (1982) 524–529.
[7] Riva E., Gerna M., Latini R., Giani P. et al.: Pharmacokinetics of amiodarone in man. J. Cardiovasc. Pharmacol. 4 (1982) 264–269.
[8] Holt D. W., Tucker G. T., Jackson P. R., Storey G. C. A.: Amiodarone pharmacokinetics. Am. Heart J. 106 (1983) 840–847.
[9] Plomp T. A., Van Rossum J. M., Robles de Medina E. O., Van Lier T., Maes R. A. A.: Pharmacokinetics and body distribution of amiodarone in man. Arzneim. Forsch. Drug Res. 34 (1984) 513–520.
[10] Berger Y., Matteazzi J. R., Pacco M.: The pharmacokinetic profile of amiodarone in man after single intravenous administration. Internal Report, Sanofi, L 3428 PS013/02 (1983).
[11] Pourbaix S., Berger Y., Desager J. P., Pacco M., Harvengt C.: Absolute bioavailability of amiodarone in normal subjects. Clin. Pharmacol. Ther. 37 (1985) 118–123.
[12] Nolan P. E., Mayersohn M., Fenster P. E., Bliss M.: Single-dose pharmacokinetics of amiodarone. Drug Intell. Clin. Pharm. 19 (1985) 463.
[13] Tucker G. T., Jackson P. R., Storey G. C. A., Holt D. W.: Bioavailability of amiodarone. Eur. J. Clin. Pharmacol. 26 (1984) 533–534.
[14] Flanagan R. J., Storey G. C. A., Holt D. W., Farmer P. B.: Identification and measurement of desethylamiodarone in blood-plasma specimens from amiodarone-treated patients. J. Pharm. Pharmacol. 34 (1982) 638–643.
[15] Latini R., Reginato R., Burlingame A. L., Kates R. E.: High-performance liquid chromatographic isolation and fast atom bombardment mass spectrometic identification of di-N-desethylamiodarone, a new metabolite of amiodarone in the dog. Biomed. Mass Spectrom. 11 (1984) 466–471.
[16] Staiger C., Jauernig R., Vries J. D., Weber E.: Influence of amiodarone on antipyrine pharmacokinetics in three patients with ventricular tachycardia. Br. J. Clin. Pharmacol. 18 (1984) 263.
[17] Harris L., Hind C. R. K., McKenna W. J., Savage C., Krikler S., Storey G. C. A., Holt D. W.: Renal elimination of amiodarone and its desethyl metabolite. Postgrad. Med. J. 59 (1983) 442–444.
[18] Bonati M., Galletti F., Volpi A., Cumetti C., Tognoni G.: Amiodarone in patients on long-term dialysis. N. Engl. J. Med. 308 (1983) 906.
[19] Adams P. C., Holt D. W., Storey G. C. A., Morley A. R., Campbell R. W. F.: Amiodarone and its desethyl metabolite: tissue distribution and morphological changes during chronic therapy. Circulation 72 (1985) 1064–1075.
[20] Rosenbaum M. B., Chaile P. A., Halpern M. S., Nau G. J. et al.: Clinical efficacy of amiodarone as an antiarrhythmic agent. Am. J. Cardiol. 38 (1976) 934–944.

[21] Lalloz M. R. A ., Byfield P. G. H., Greenwood R. M., Himsworth R. L.: Binding of amiodarone by serum proteins and the effects of drugs, hormones and other interacting ligands. J. Pharm. Pharmacol. 36 (1984) 366–372.
[22] Holt D. W., Storey G. C. A.: Amiodarone pharmacokinetics. In: New Aspects in the Medical Treatment of Tachyarrhythmias. Eds.: Breithardt G., Loogan F. Urban & Schwarzenberg, Munich 1983, pp. 69–72.
[23] McKenna W. J., Harris L., Rowland E., Whitelaw A., Storey G. C. A., Holt D. W.: Amiodarone therapy during pregnancy. Am. J. Cardiol. 51 (1983) 1231–1233.
[24] Pitcher D., Leather H. M., Storey G. C. A., Holt D. W.: Amiodarone in pregnancy. Lancet 1 (1983) 597–598.
[25] Robson D. J., Raj M. V. J., Storey G. C. A., Holt D. W.: Use of amiodarone during pregnancy. Postgrad. Med. J. 61 (1985) 75–77.
[26] Harris L., McKenna W. J., Rowland E., Holt D. W., Krikler D. M.: Amiodarone: side-effects of long-term therapy. Circulation 67 (1983) 45–51.
[27] Zachary C. B., Slater D. N., Holt D. W., Storey G. C. A., Macdonald D. M.: The pathogenesis of amiodarone-induced pigmentation and photosensitivity. Br. J. Dermatol. 110 (1984) 451–456.
[28] Lim P. K., Trewby P. N., Storey G. C. A., Holt D. W.: Neuropathy and fatal hepatitis in a patient receiving amiodarone. Br. Med. J. 288 (1984) 1638–1639.
[29] Marcus F. I.: Drug interactions with amiodarone. Am. Heart J. 106 (1983) 924–930.
[30] Hamer A., Peter T., Mandel W. J., Scheinman M. M., Weiss D.: The potentiation of warfarin anticoagulation by amiodarone. Circulation 65 (1982) 1025–1029.
[31] Almog S., Shafron N., Halkin H.: Mechanism of warfarin potentiation by amiodarone: dose and concentration dependent inhibition of warfarin elimination. Eur. J. Clin. Pharmacol. 28 (1985) 257–261.
[32] Moysey J. O., Jaggarao N. S. V., Grundy E. W. , Chamberlain D. A.: Amiodarone increases plasma digoxin concentrations. Br. Med. J. 282 (1981) 272.
[33] Nademanee K., Kannan R., Hendrickson J., Ookhtens M. et al.: Amiodarone-digoxin interaction: clinical significance, time course of development, potential pharmacokinetic mechanisms and therapeutic implications. J. Am. Coll. Cardiol. 4 (1984) 111–116.
[34] Koren G., Hesslein P. S., MacLeod S. M.: Digoxin toxicity associated with amiodarone therapy in children. J. Pediatrics 104 (1984) 467–470.
[35] Fenster P. E., White N. W., Hanson C. D.: Pharmacokinetic evaluation of the digoxin-amiodarone interaction. J. Am. Coll. Cardiol. 5 (1985) 108–112.

[36] Robinson K. C., Johnston A., Walker S., Mulrow J. P., McKenna W. J., Holt D. W.: The digoxin-amiodarone interaction. Cardiovasc. Drug Ther. 3 (1989) 25–28.

[37] McKenna W. J., Rowland E., Krikler D. M.: Amiodarone: the experience of the past decade. Br. Med. J. 287 (1983) 1654–1656.

[38] Mostow N. D., Rakita L., Vrobel T. R., Noon D., Blumer J.: Amiodarone: intravenous loading for suppression of complex ventricular arrhythmias. JACC 4 (1984) 97–104.

[39] Staubli M., Bircher J., Galeazzi R. L., Remund H., Studer H.: Serum concentrations of amiodarone during long term therapy. Eur. J. Clin. Pharmacol. 24 (1983) 485–494.

[40] Fraser A. G., Stephens M. R., Newcombe R. G., Holt D. W.: Association of serum desethylamiodarone concentration with adverse effects of amiodarone. Br. J. Clin. Pharmacol. 18 (1984) 276.

[41] Adams P. C., Sloan P., Morley A. R., Holt D. W.: Peripheral neutrophil inclusions in amiodarone-treated patients. Br. J. Clin. Pharmacol. 22 (1986) 736–738.

3 Management of Patients on Amiodarone

William J. McKenna

3.1 Introduction

Side effects are common during amiodarone therapy and often limit use of the drug [1, 2]. Some side effects are idiosyncratic, others occur in patients with hepatic, pulmonary or thyroid disease which predisposes them to complications of therapy [3]. Many of the troublesome as well as the serious side effects of amiodarone, however, are related to the dose and duration of therapy [4]. When long-term use of amiodarone is necessary it is logical to use the drug in the lowest effective dose. Though the recommended maintenance dose of 200–400 mg daily is often effective, this may result in higher plasma concentrations than are necessary; the problem is how to achieve the minimum effective dose. There is a linear relationship of dose and plasma concentration of both amiodarone and

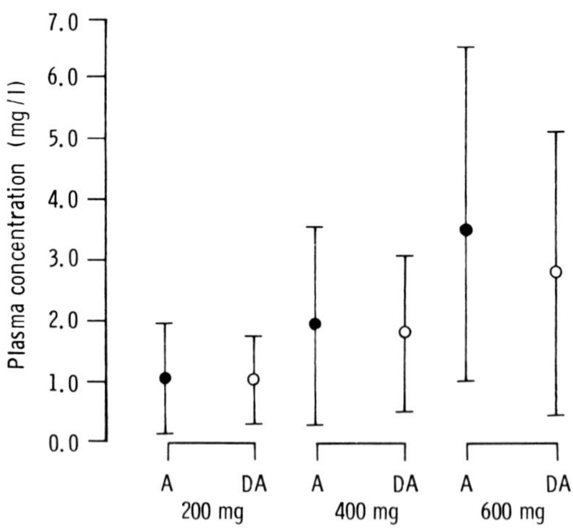

Fig 3.1: Mean ±2SD plasma concentrations of amiodarone and desethylamiodarone with respect to daily dosage of amiodarone.

the metabolite desethylamiodarone, but for a given dose there is an extremely wide range of plasma concentrations (fig 3.1), in part due to the variable oral bioavailability [5]. It is therefore important not only to have a reliable assessment of efficacy but also to measure plasma concentrations of amiodarone and desethylamiodarone as a guide to dosage adjustment. In this manuscript is presented a protocol for the management of patients on amiodarone which has evolved in large part from the experience described below in patients with hypertrophic cardiomyopathy. The approach outlined is only feasible with arrhythmias which are not associated with ventricular fibrillation or symptoms of impaired consciousness. Even in such cases, however, the principle of achieving the lowest effective dose remains valid but the approach must be modified with in-hospital observation and programmed electrical stimulation or electrocardiographic monitoring to assess efficacy.

3.2 Patients

Fifty-three patients with hypertrophic cardiomyopathy who had serious arrhythmias (45 patients), refractory chest pain (five patients) or a high risk of sudden death (three patients) received amiodarone for six to 96 months (median 18) after completion of a loading dose (600–1,200 mg daily for one to two weeks) and an initial maintenance period (200–400 mg daily for one to two months).

Efficacy was assessed with 48 hour electrocardiographic monitoring and was defined as:

- abolition of episodes of ventricular tachycardia,
- peak ventricular extrasystolic counts of ≤100/24 hour and ≤10/hour or a 90% reduction versus baseline,
- abolition of episodes of supraventricular tachycardia or paroxysmal atrial fibrillation.

Plasma concentrations of amiodarone and desethylamiodarone were measured in samples collected at least eight hours after the last amiodarone dose, using a high performance liquid chromatographic assay [6].

3.3 Results

3.3.1 Efficacy

Between the initial maintenance period and last follow-up, 57 alterations in dosage (range 0 to seven, median two) were made in the 29 patients who received amiodarone for ventricular arrhythmias. The dose of amiodarone was increased in three patients because of failure to control arrhythmia and was decreased by 50 to 200 mg daily (median 100 mg) in 17, one to six months after the initial maintenance period. At evaluation three to six months after the dosage reduction, suppression of serious ventricular arrhythmia was maintained in 12 of 15; 13 patients in whom serious ventricular arrhythmias recurred had control restored when amiodarone was increased.

Two patients had progressive increase in the number of ventricular extrasystoles during more than three years of amiodarone therapy; the daily dose was not increased because they experienced side effects and ventricular tachycardia was absent. At last follow-up, the median dose was 300 mg/day and ventricular tachycardia was suppressed in 24 of 26 patients. The minimum effective plasma concentrations of amiodarone and desethylamiodarone are shown in fig 3.2 and were 1.35 ± 0.74 and 1.4 ± 0.67 mg/l respectively.

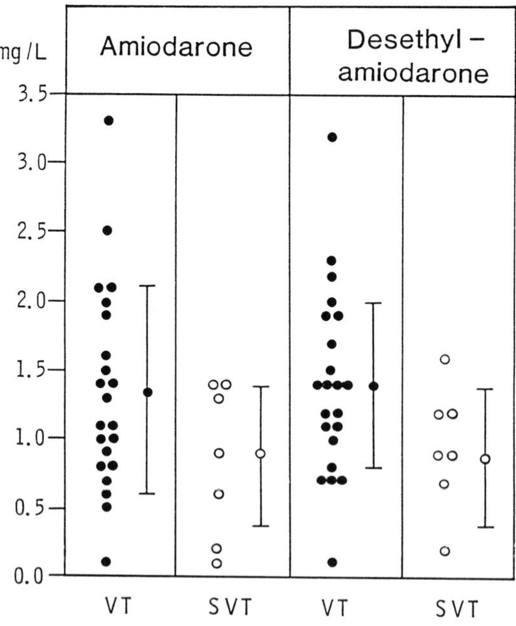

Fig 3.2: Minimum effective plasma concentrations of amiodarone (A) and desethylamiodarone (DA) in 22 patients with ventricular tachycardia (VT) and seven patients with supraventricular tachycardia (SVT).

Nine patients with hypertrophic cardiomyopathy received amiodarone for the treatment of frequent or prolonged supraventricular tachycardia which had been refractory to beta blockers and digoxin. After the initial maintenance period, the dose was reduced in eight patients and was increased in one; a further six alterations were made over 6 to 42 months (median 11) and at last follow-up arrhythmias were suppressed in eight with a median dose of 300 mg/day. The plasma concentration of amiodarone and desethylamiodarone at last follow-up was 0.88 ± 0.5 mg/l (fig 3.2). One patient had more frequent episodes of atrial flutter which became relatively refractory to D/C cardioversion and amiodarone was discontinued.

3.3.2 Unwanted Effects of Amiodarone

During the loading and the initial maintenance period, sleep disturbance was the most common unwanted effect (fig 3.3); 14 (26%) patients experienced vivid dreams, nightmares and/or early morning waking with difficulty in getting back to sleep. This resolved in seven patients in association with dose reduction and

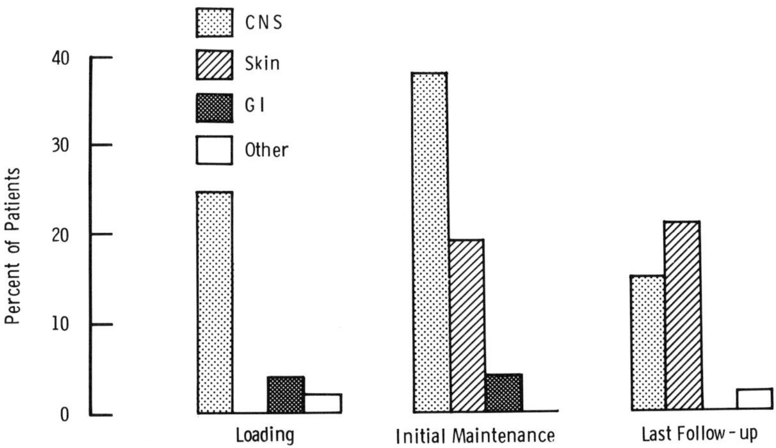

Fig 3.3: Central nervous system (CNS) side effects of sleep disturbances, tremor and headaches were common during the loading (25%) and initial maintenance period (38%), but responded to reduction of the dose of amiodarone. Skin photosensitivity was not reported until after the initial maintenance period and did not respond to a reduction of the dose. Gastrointestinal complaints (GI) were rare and were only seen during the loading and initial maintenance periods, particularly in the elderly.

in one patient who remained on 400 mg daily. Of six patients who still had sleep disturbance at last follow-up three were receiving 200 mg and the remainder 400 mg daily.

Dermatological side effects were also common. 41 patients (77%) reported photosensitivity; in 11 (21%) this was particularly troublesome and required application of barrier creams and avoidance of exposure to sunlight. In these patients, the daily dose was 100–400 mg (median 300) and dose reduction did not reduce the photosensitivity. Two patients developed blue/grey facial discoloration on amiodarone. One received a loading dose of 1,200 mg of amiodarone for one week, following which there was near total suppression of both ventricular and supraventricular arrhythmia. He then received 400 mg of amiodarone daily for the following month, however, his arrhythmias recurred. The daily amiodarone dose was increased to 600 mg and on this regimen the arrhythmias were suppressed; one and a half years later he developed blue facial discoloration. When amiodarone was discontinued, plasma amiodarone and desethylamiodarone levels were 3.2 and 2.5 mg/l respectively; ventricular arrhythmias were suppressed for approximately 20 weeks without additional therapy (fig. 3.4). The other patient who developed facial discoloration had experienced mild photosensitivity. He received a maintenance dose of 300 mg daily for five years for control of refractory atrial fibrillation; plasma amiodarone and desethylamiodarone concentrations were 2.2 and 2.3 mg/l respectively. His symptomatic improvement was such that he preferred to continue amiodarone.

One patient developed dry cough dyspnoea with diffuse shadowing on chest X-ray (fig 3.5). He received 200 mg daily from February 1980 until March 1983, an approximate total dose of 110 grams. Plasma amiodarone and desethylamiodarone concentrations were 0.9 and 0.8 mg/l. Open lung biopsy revealed patchy areas of fibrosis within normal lung parenchyma. There were no lymphocytes, eosinophils or evidence of inflammatory cell reaction. Steroid therapy was continued for two years because serial bronchial washings revealed the appearance of eosinophils after symptoms and the radiological abnormalities had improved.

Four patients complained of tremor during the loading period; in two, this was noted as an exacerbation of a previously noticeable familial tremor. All four responded within three to six days to dose reduction and one of the patients with familial tremor was further improved with propranolol (40 mg daily). Another patient who received maintenance therapy of 200 mg daily for one year experienced tremor in association with the sensation of being unsteady on her feet and an overwhelming and persistent feeling of anxiety. Apart from the tremor, neurological investigations revealed no abnormalities. After amiodarone was discontinued, these symptoms, including the tremor, were unchanged for over two weeks before they gradually resolved. Control of ventricular and supra-

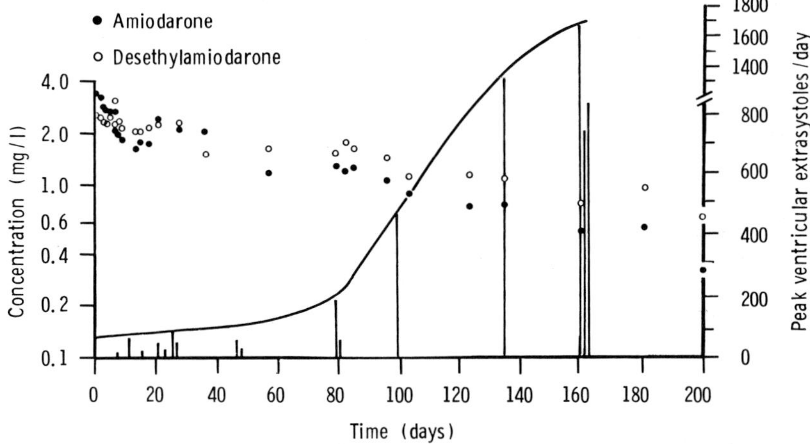

Fig 3.4: The relation of the plasma concentration of amiodarone and desethylamiodarone and peak daily ventricular extrasystoles after amiodarone was discontinued. The plasma concentrations are plotted on a log-linear scale in relation to days after stopping amiodarone. The number of ventricular extrasystoles detected during 24 hour electrocardiographic monitoring is superimposed; these arrhythmias were controlled for approximately 20 weeks, when the plasma concentration of amiodarone and desethylamiodarone was 0.75 and 1.0 mg/l, respectively.

ventricular tachycardia has since been maintained on 600 mg a week without side effects.

During follow-up transient proximal limb girdle myopathy, conjunctivitis sicca, nausea (which responded poorly to dose reduction) and global alopecia were reported. One patient, who was clinically euthyroid but had a TSH >25 mU/l prior to amiodarone, became hypothyroid and required thyroid replacement. Nine (17%) had increased total T4; 17 (32%) had transient elevation of hepatic transaminases. None had neuropathy, visual symptoms, clinical hyperthyroidism or clinical hepatic disease.

3.4 Discussion

Our experience of amiodarone in hypertrophic cardiomyopathy was characterised by control of arrhythmia and a high incidence of sleep disturbance and photosensitivity. Serious side effects were rare; amiodarone was withdrawn in four of 53 patients and in two of these it was subsequently restarted because arrhythmias could not be controlled by other means. Peripheral neuropathy and

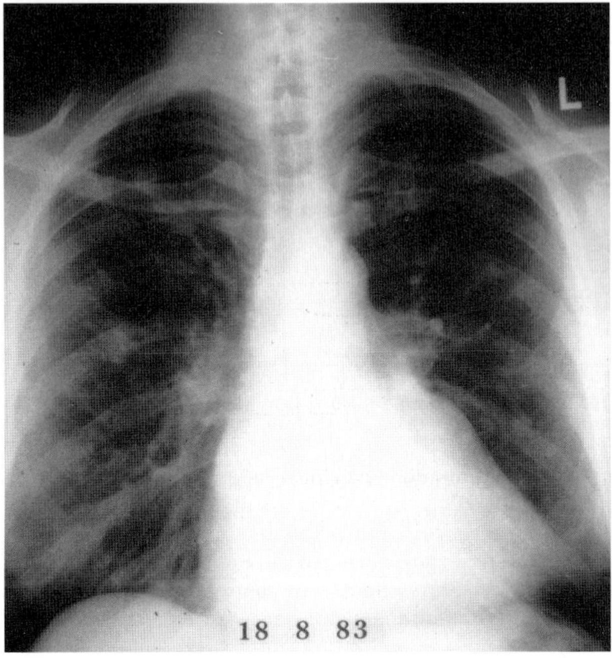

Fig 3.5: Chest x-ray of a 65 year old man who developed diffuse pulmonary shadowing following long term low dose amiodarone. The persistence of inflammatory changes long after plasma levels of amiodarone were undetectable suggested an immune mediated rather than a direct toxic effect of amiodarone.

exacerbation of ventricular arrhythmia were not seen and the most serious complication observed was blue-grey facial discoloration in two patients and pulmonary fibrosis in another. This contrasts with recent reports of a 15 to 30% incidence of serious and life threatening side effects in patients receiving at least 400 mg daily for one to 41 months [1, 2]. The majority of the unwanted effects of amiodarone are dose/duration dependent. The rarity of serious complications in patients who received modest doses (median 300 mg daily) underscores the importance of administering the lowest effective dose.

These patients were started on amiodarone between 1978 and 1982 when different loading regimens were used. An initial dose of 600 or 800 mg daily for one week was most practical for achieving early control of arrhythmia without side effects. This policy evolved from our experience and does not reflect the results of a prospective comparative study of different loading regimens. The original decision to evaluate patients after the initial maintenance period (i.e., after five to nine weeks of treatment) was pragmatic, but pharmacokinetic data indicate that this is appropriate. At last follow-up, the median dose of amiodarone was 300 mg for those with ventricular arrhythmia, paroxysmal supraventricular arrhythmia and atrial fibrillation. This experience of loading and maintenance dosage represents a balance between arrhythmia suppression and side effects that was practical and effective in patients with hypertrophic cardiomyopathy. The principle of achieving a minimum effective dose is valid; however, the specific doses cannot be extrapolated to the treatment of arrhythmias associated with other conditions.

Three patients received 600 mg daily, long term; as ventricular arrhythmias were not controlled after the initial maintenance period, this was considered necessary and no attempt was made to reduce the dose. Plasma levels of amiodarone and desethylamiodarone were above the therapeutic range and arrhythmia was suppressed for 20 weeks after stopping treatment; in retrospect, it is probable that one of these patients would have been effectively controlled at lower plasma levels, which were observed in the other patients with ventricular arrhythmia (fig 3.2). Measurement of plasma drug concentration at an earlier stage in his treatment (such measurements were not routinely available in 1980 when he started treatment) would have alerted us to the high probability of toxicity and to the possibility that he was receiving more amiodarone than necessary. With the experience of such cases and the recognition of the high incidence of dose-related side effects, we now attempt to achieve the lowest effective maintenance dose whenever this is feasible. Our protocol in this regard is detailed below.

3.5 Protocol

After the exclusion of diseases which may be complicated by amiodarone therapy, the drug is administered as a loading dose, usually 600–800 mg daily for five to seven days, followed by an initial maintenance dose of 400 mg daily for the next one to two months. The patient should be evaluated after the initial maintenance period and at three to six month intervals with assessment of side effects, assessment of efficacy from symptoms, 24–48 hour electrocardiographic monitoring or programmed electrical stimulation as appropriate, and measurement of plasma amiodarone and desethylamiodarone drug concentrations. The daily dose of amiodarone should then be decreased or increased by 50–200 mg in response to the absence or presence of arrhythmias. The magnitude of the dosage change is guided by side effects and plasma amiodarone and desethylamiodarone levels. When arrhythmia is controlled without side effects, a plasma concentration of 0.5 mg/l is considered to be the minimum effective level below which the daily dose is not reduced. When arrhythmia is not controlled with a plasma concentration of amiodarone of 1.5 mg/l or less, an additional drug is added rather than increase the daily dose when long term treatment is required, as such plasma concentrations are more frequently associated with the development of serious side effects.

Acknowledgements

This paper is based on data previously published in McKenna W. J., Harris L., Rowland E., et al.: Amiodarone for long-term management of patients with hypertrophic cardiomyopathy. Am. J. Cardiol. 54 (1984) 802–810.

3.6 References

[1] McGovern B., Garan H., Kelly E., Ruskin J. N.: Adverse reactions during treatment with amiodarone hydrochloride. Br. Med. J. 287 (1983) 175–180.

[2] Fogoros R. N., Anderson K. P., Winkle R. A., Swerdlow C. D., Mason J. W.: Amiodarone: clinical efficacy and toxicitiy in 96 patients with recurrent drug-refractory arrhythmias. Circulation 68 (1983) 88–94.

[3] McKenna W. J., Rowland E., Krikler D. M.: Amiodarone: the experience of the past decade. Br. Med. J. 287 (1983) 1654–1656.

[4] Harris L., McKenna W. J., Rowland E., Holt D. W., Storey G. C. A., Krikler D. M.: Side effects of long term amiodarone therapy. Circulation 67 (1983) 45–51.

[5] Holt D. W., Tucker G. T., McKenna W. J.: Amiodarone pharmacokinetics. Br. J. Clin. Prac. 40 (suppl. 44) (1986) 109–114.

[6] Storey G. C. A., Holt D. W.: High-performance liquid chromatographic measurement of amiodarone and its desethyl metabolite in plasma or serum at the concentrations attained following a single 400 mg dose. J. Chromatogr. 245 (1982) 377–380.

Publications of the DFG Senate Commission

Mitteilung I	Gaschromatographische Retentionsindices toxikologisch relevanter Verbindungen	1982
Report II	Gas-Chromatographic Retention Indices of Toxicologically Relevant Substances on SE-30 or OV-1 (Second, Revised and Enlarged Edition)	1985
Mitteilung III	Empfehlungen zum Nachweis von Suchtmitteln im Urin	1985
Report IV	Pharmacokinetics: Classic and Modern – Application to Pharmacology and Toxicology	1985
Report V	Theophylline Profile	1985
Mitteilung VI	Dünnschichtchromatographische Suchanalyse für 1,4-Benzodiazepine in Harn, Blut und Mageninhalt	1986
Report VII	Thin-Layer Chromatographic R_f-Values of Toxicologically Relevant Substances on Standardized Systems	1987
Mitteilung VIII	Photometrische Bestimmung von Carboxy-Hämoglobin (CO-Hb) im Blut	1988
Report IX	Tobramycin Profile	1988

Mitteilung X	Empfehlungen zur klinisch-toxikologischen Analytik, Folge 1: Einsatz von immunochemischen Testen in der Suchtmittelanalytik	1988
Mitteilung XI	Empfehlungen zur klinisch-toxikologischen Analytik, Folge 2: Einsatz der Gaschromatographie in der klinisch-toxikologischen Analytik	1988
Mitteilung XII	Empfehlungen zur klinisch-toxikologischen Analytik, Folge 3: Einsatz der Hochleistungsflüssigchromatographie in der klinisch-toxikologischen Analytik	1989
Denkschrift	Klinisch-toxikologische Analytik	1983
Denkschrift	Dokumentation und Information in der klinisch-toxikologischen Analytik	1987
Rundgespräche und Kolloquien	Klinisch-toxikologische Analytik – Gegenwärtiger Stand und Forderungen für die Zukunft	1987
Rundgespräche und Kolloquien	Klinisch-toxikologische Analytik bei akuten Vergiftungen und Drogenmißbrauch	1989

Addresses of the Authors

Dr. David W. Holt	Director The Analytical Unit St. George's Hospital Medical School London SW17 ORE England
Prof. Dr. Robert A. A. Maes	NIDDR – Nederlands Instituut voor Drugs en Doping Research Vondellaan 14 NL-3521 GE Utrecht
Dr. William J. McKenna	Department of Cardiological Sciences St. George's Hospital Medical School Cranmer Terrace London SW17 ORE England
Dr. Theo A. Plomp	Rijksuniversiteit Utrecht Toxicologisch Centrum Vondellaan 14 NL-3521 GE Utrecht

Members of the DFG Senate Commission for Clinical-Toxicological Analysis

Dr. J. Bäumler	Sonnmattstraße 20 CH-4142 Münchenstein
Prof. Dr. H. Brandenberger (Member until summer 1987, later as guest)	Lindenhofrain 8 CH-8708 Männedorf und Universität Zürich CH-8006 Zürich
Prof. Dr. Dr. J. Büttner	Institut für Klinische Chemie I Zentrum Laboratoriumsmedizin Medizinische Hochschule Hannover Konstanty-Gutschow-Straße 8 D-3000 Hannover 61
Prof. Dr. M. von Clarmann	Toxikologische Abteilung II. Medizinische Klinik und Poliklinik Technische Universität Ismaninger Straße 22 D-8000 München 80
Prof. Dr. Dr. M. Geldmacher- von Mallinckrodt (Chairperson)	Institut für Rechtsmedizin Universitätsstraße 22 D-8520 Erlangen
Prim. Dr. H. J. Gibitz	Chemische Zentrallaboratorien Landeskrankenhaus Salzburg Müllner Hauptstraße 48 A-5020 Salzburg
Prof. Dr. A. N. P. van Heijst (Member until summer 1987, later as guest)	Baarnseweg 42A NL-3735 MJ Bosch en Duin

Prof. Dr. K. Ibe	Freie Universität Berlin Universitätsklinikum Rudolf Virchow Standort Charlottenburg Spandauer Damm 130 D-1000 Berlin 19
Prof. Dr. G. Machata	Chemische Abteilung Institut für Gerichtliche Medizin Sensengasse 2 A-1090 Wien IX
Prof. Dr. R. A. A. Maes	NIDDR – Nederlands Instituut voor Drugs en Doping Research Vondellaan 14 NL-3521 GE Utrecht
Dr. H. Moll	Kinderklinik des Marienhospitals Hauptkanal 75 D-2990 Papenburg
Prof. Dr. M. Oellerich	Institut für Klinische Chemie I Zentrum Laboratoriumsmedizin Medizinische Hochschule Hannover Konstanty-Gutschow-Straße 8 D-3000 Hannover 61
Prof. Dr. H. Schütz	Institut für Rechtsmedizin Frankfurter Straße 58 D-6300 Gießen
Prof. Dr. Dr. D. Stamm	Abt. für Klinische Chemie Max-Planck-Institut für Psychiatrie Kraepelinstraße 10 D-8000 München 80
Dr. M. Stoeppler	Institut für Chemie Forschungszentrum GmbH Postfach 1913 D-5170 Jülich 1

Prof. Dr. R. Wennig	Laboratoire National de Santé Boîte postale 1102 1 A, rue Auguste Lumière L-1011 Luxembourg
Prof. Dr. Dr. H. Wisser	Abteilung für Klinische Chemie Robert-Bosch-Krankenhaus Auerbachstraße 10 D-7000 Stuttgart 50
Prof. Dr. R. A. de Zeeuw	Rijksuniversiteit Groningen Vakgroep Analytische Chemie en Toxicologie Antonius Deusinglaan 2 NL-9713 AW Groningen

Permanent Guest:

Dr. W. Fabricius	Bundesgesundheitsamt Thielallee 88–92 D-1000 Berlin 33